THE CORONA VISION

by

a Friend of Medjugorje

THE CORONA VISION

©2020 S.J.P Lic. to:
CARITAS OF BIRMINGHAM
STERRETT, ALABAMA 35147 USA

About the Witness

Many who will read these books have been following the writings of a Friend of Medjugorje for years. His original and unique insights into the important events of our day have won credence in hundreds of thousands of hearts around the world, with those affecting others, thereby, touching into the millions. His moral courage in the face of so many leaders caving in to the pressures of a politically correct world is not only refreshing, but, according to tens of thousands of written testimonies over 32 years, has helped to strengthen deeply those who desire to live the fullness of their Christian faith. His insights have repeatedly proven prophetic, having their source in the apparitions of the Virgin Mary in Medjugorje. Deeply and personally influenced by the events surrounding Medjugorje, he gave himself to the prayerful application of the words of the Virgin Mary into his life. He has spoken all over the world on Our Lady's messages and how to put them into everyday life. He came to understand that Our Lady was sent by God to speak to mankind in this time because the dangers man is facing are on a scale unlike any the world has ever known since Noah and the flood. He is not an author. He is a witness of what Our Lady has shown him to testify to—first, by his life—secondly, through the written word. He is not one looking in from the outside regarding Medjugorje, but one who is close to the events—many times, right in the middle of the events about which he has written.

Originally writing to only a few individuals in 1987, readership has grown well into the millions in the United States and in over 130 foreign countries, who follow the spiritual insights and direction given through these writings.

When asked why he signs only as "a Friend of Medjugorje," he stated:

"I have never had an ambition or desire to write. I do so only because God has shown me, through prayer, that He desires this of me. So from the beginning, when I was writing to only a few people, I prayed to God and promised I would not sign anything; that the writings would have to carry themselves and not

be built on a personality. I prayed that if it was God's desire for these writings to be inspired and known, then He could do it by His Will and grace and that my will be abandoned to it.

*"The Father has made these writings known and continues to spread them to the ends of the earth. These were Our Lord's last words before ascending: **'Be a witness to the ends of the earth.'** These writings give testimony to that desire of Our Lord, to be a witness with one's life. It is not important to be known. It is important to do God's Will."*

For those who require "ownership" of these writings by the 'witness' in seeing his name printed on this work in order to give it more credibility, we, Caritas of Birmingham and the Community of Caritas, state that we cannot reconcile the fact that these writings are producing hundreds of thousands of conversions, easily will be into the millions, through God's grace. His writings are requested worldwide from every corner of the earth. His witness and testimony, therefore, will not take credit for a work that, by proof of the impact these writings have to lead hearts to conversion, have been Spirit-inspired, with numbers increasing yearly, sweeping as a wave across the ocean. Indeed, in this case, crossing every ocean of the earth. Our Lady gave this Witness a direct message for him, through the Medjugorje visionary, Marija, which part of what Our Lady said to him was to **"…witness not with words but through humility…"** (Oct. 6, 1986) It is for this reason that he wishes to remain simply "A Friend of Medjugorje."

In order to silence the voice of this witness, darkness has continually spewed out slanders to prevent souls from reading his convicting and life-changing writings. For if these writings were not so, darkness would ignore or even lead people to them. But Jesus promised persecution to all those who follow Him, and the same will be to those who follow His Mother. *"If they persecuted me, they will also persecute you."* John 15:20

As a witness in real-time of Our Lady's time on earth, his witness and writings will continue to speak—voicing Our Lady's Way to hundreds of millions not yet born—in the centuries to come.

— Caritas of Birmingham

Medjugorje
The Story in Brief

A VILLAGE SEES THE LIGHT is the title of a story which "Reader's Digest" published in February 1986. It was the first major news on a mass public scale that told of the Virgin Mary visiting the tiny village of Medjugorje, Bosnia-Hercegovina. At that time this village was populated by 400 families.

It was June 24, 1981, the Feast of John the Baptist, the proclaimer of the coming Messiah. In the evening, around 5:00 p.m., the Virgin Mary appeared to two young people, Mirjana Dragičević* and Ivanka Ivanković*. Around 6:40 p.m. the same day, Mirjana and Ivanka, along with four more young people, Milka Pavlović*, the little sister of Marija, Ivan Ivanković, Vicka Ivanković*, and Ivan Dragičević saw the Virgin Mary. The next day, June 25, 1981, along with Mirjana, Ivanka, Vicka and Ivan Dragičević, Marija Pavlović* and Jakov Čolo also saw the Virgin Mary, bringing the total to six visionaries. Milka Pavlović* and Ivan Ivanković only saw Our Lady once, on that first day. These six have become known as and remain "the visionaries."

These visionaries are not related to one another. Three of the six visionaries no longer see Our Lady on a

* Names at the time of the apparitions, they are now married with last names changed.

daily basis. As of April 2020, the Virgin is still appearing everyday to the remaining three visionaries; that's well over 16,705 apparitions. This count is each day for all the visionaries together in the apparitions. The visionaries have been separated for more years than together, which means the number is minimum 30 years × 3 visionaries who still see Our Lady daily being separated during apparition time.

The supernatural event has survived all efforts of the Communists to put a stop to it, many scientific studies, and even the condemnation by the local bishop; yet, the apparitions have survived, giving strong evidence that this is from God because nothing and no one has been able to stop it. For over 39 years, the apparitions have proved themselves over and over and now credibility is so favorable around the world that the burden of proof that this is authentic has shifted from those who believe to the burden of proof that it is not happening by those opposed to it. Those against the apparitions are being crushed by the fruits of Medjugorje — millions and millions of conversions which are so powerful that they are changing and will continue to change the whole face of the earth.

See **mej.com** for more information.
or **Medjugorje.com**

The Corona Vision

by a Friend of Medjugorje

Through these 39 years of Our Lady's apparitions in Medjugorje, no one has understood, let alone promoted, a profound Medjugorje prophecy for the future. Only through the mission of Caritas of Birmingham have you heard of it, through the years. In the Book of Daniel, Nebuchadnezzar had a powerful dream; a vision. Though he saw, he did not understand what it meant. Nebuchadnezzar called upon Daniel to interpret what he experienced in his vision.

In the beginning days of Medjugorje, Our Lady showed a very important window into the future. Those who were given the grace to "look through the window," could not, however, tell why or for what purpose the huge vision was given. Even to this most recent date, April 2020, its meaning has not been understood. Why? God does not give prophetic signs for all to understand until they are fulfilled. Even the prophets do not always fully understand their inspirations, spiritual experiences, etc. In <u>The Poem of the Man-God</u>,[1] Jesus was asked by His apostles, "Did Joachim and Anne* know Mary was the chosen Virgin?" Jesus said: "They did not know." But Jesus said that Joachim intuited that little Mary was "Immaculate." Jesus, explaining further, said:

> *"Joachim spoke, inspired by God, like the prophets. He did not understand the supernatural truth that the spirit spoke through his lips."* (Jesus continued) *"Joachim was a prophet. His soul repeated like an echo what God said to his soul."*[**]

To come to an understanding of God's inspirations takes daily conversion, prayer, fasting, cogni-

---

* Joachim and Anne were the Blessed Virgin Mary's parents.
** <u>The Poem of the Man-God</u> Vol. 2, by Maria Valtora, p. 288.

tion gained through life experiences, being just and humble, and for our times, not only being immersed in all of Our Lady's messages, but most importantly, being obedient to what She is saying, both in small and large things.

Many have the possibility of being a prophet, but God can only use those who open their hearts, pray and fast, and have no self-interest outside of God's plans. It is to these souls that God gives His illuminations and only then, one can grasp a vision and define it. The prophets were persecuted and, eventually, many were put to death. There are those who are in fear of losing their positions of being "in the circle." This is why many are not given understanding of Medjugorje's broadness, its breadth and depth. Medjugorje has no boundaries. Why would God give understanding of the mysteries of Medjugorje, of that which is to be revealed, to those who would allow themselves to be silenced out of fear of a loss of reputation, position or even one's life? Authorities in Medjugorje have suppressed and persecuted those who dare to step out and speak about what they, "the Authorities," have censored.

Fr. Petar Ljubicic was chosen by Medjugorje visionary, Mirjana, to read and reveal the Secrets, once Our Lady indicates to Mirjana the time of their release has come. Pause and think about it for a moment. Do you think he is, just easily, going to read them to the world? There is no possibility that he will be able to read the Secrets without the devil working through the authority who is over Fr. Petar, in order to stop him from reading them. You can take that "common sense" statement "prophetically." Our Lady said:

December 2, 2007

> **"...'God's Word' which is the light of salvation and 'the light of common' sense."**

Jesus said certain demons can only be cast away — blocked out of the way — by fasting. Does it take that much effort to read a secret? If you are given a secret to read, you just read it. It is that simple. But, if that Secret is going to prove the apparitions of the Queen of Peace of Medjugorje to the whole world, and thereby, change the world, the devil, with all of hell, will use all his power to stop the Secrets from being read. How? By the devil acting through anyone who has the ability or authority to stop Fr. Petar.

Through Church history, we know what superiors have done with many of the saints who were given important missions to fulfill. Often missions were then hampered, oppressed, and in some cases, even destroyed by their superiors or other authorities. Knowing this, Fr. Petar will not just read the Secret(s) without monumental satanic interference against him. If you think he will be able to, then why is the Virgin Mary requiring Fr. Petar and Mirjana to fast on bread and water for 7 days* before he is to read a simple statement that will happen three days after he reads it? Remember what Jesus said to the apostles

---

* Fr. Petar is to read the parchment the Virgin Mary gave directly to Mirjana with the Secrets written on it. Fr. Petar will receive it 10 days before the Secret happens. He is to fast 7 days. Then 3 days before the Secret happens, Fr. Petar will announce to the public the Secrets, of what will happen before it happens.

after they were trounced by the demons, when the apostles tried to drive the demons away? The demons turned out to be more than what the apostles could handle and the apostles were humiliated by their failure to cast the demons out. Jesus told the apostles these demons require fasting and prayer to be driven away.* Fr. Petar will have to fast to stand up against and drive away the hordes of demons, who will do everything, even murder, to stop the Secrets from being read. Murder? You think not? If so, you do not understand Medjugorje. Murder is not only the killing of the body, but of one's reputation. Scripture relays that to destroy one's reputation is to murder. But even physical murder is a desire of the devil. Our Lady's fresh words of March 25, 2020, confirm this point.

> **"...satan is reigning and wants to destroy your lives and the earth on which you walk..."**

Yes, these are strong statements concerning Medjugorje, and what is written is meant as stated.

* Mark 9:28

Medjugorje is not a fragile, feel-good playground. It is a battleground between Heaven and hell, between Jesus and satan. In the physical realm, it is a battle between good vs. evil—with the people of the earth in between. Our Lady said:

January 28, 1987

> "...Whenever I come to you my Son comes with me, but so does satan..."

Medjugorje has its Pharisees, its Sadducees, its corruptors, its liars, its power grabbers and those with personal agendas. But it also has its meek, its humble, its saints, the salt of the earth people, those with no self-interest, who are wholly committed to Our Lady's Immaculate Heart, sacrificing their lives for the salvation of mankind. There are those good villagers who have been loyal to Our Lady for 39 years.

But even with the latter, good group, almost no one is willing to step out of line against the first group mentioned. It is the same erred mentality that people possessed 2,000 years ago in Jerusalem. There were many who thought and believed it was they (Pharisees, etc.) who were the saviors of their people in Jerusalem, and not the Christ. Yes, Our Lady and Medjugorje have many of the same negative dynamics that Jesus had against Him. This shouldn't surprise anyone. These souls are very cunning, giving an appearance that they are for Our Lady, but in reality they are for themselves, their agenda, their plans, and their preservation. They go way back. They say that they see, but they are blind. They have not and do not take Our Lady's messages seriously, and even reject them as inconsequential, therefore, nonessential.

Our Lady said on June 25, 1991:

> **"…There are '<u>many</u>' people who do not desire to understand my messages and to accept with seriousness what I am saying. But you I therefore call and ask that by your lives and your daily living you witness 'my presence'…"**

We <u>must</u> take all the messages Our Lady has given these 39 years (as of 2020) seriously. In addition, even the apparitions in which Our Lady gave no message are important. Marija, many times has said after an apparition, when Our Lady gives no message, that Our Lady's presence 'is' the message — **"my presence"** — as Our Lady called it in the above message (**"witness '<u>my presence</u>'"**). Also significant is every action of Our Lady, no matter how insignificant it may seem. All of Our Lady's actions are to be accepted with seriousness. You must combine together all the messages, all Our Lady's actions, Her apparitions, what She has shown us, and implement everything, <u>now</u>, in your lives. **If you do not take the messages and Our Lady's presence seriously, you will not be ready.** It is explicitly clear that Our Lady demands of us, Her apostles, to take every word of Hers and turn to them very seriously to be ready for the future. Ready for what in the future? **A new way of life in a new time, a great awakening, a great time of evangelization sweeping across the whole world.** When? When the Secrets begin and the Post Secrets Period. However, one can have now what the "great time" will deliver. It is available now, if you act with seriousness to Her call.

When the time comes for Fr. Petar to read the Secrets, the whole Medjugorje world not only should, but <u>must</u> fast with Fr. Petar for seven days for the grace for the Secrets to be read publicly, without delay and that they be read in their fullness. Our Lady has relayed to us that even if God has decided that an intention is going to be granted, the more the prayer attached to it, greater will be the graces He will give.

Our Lady chose this mission of Caritas and grew it to where it is today. From the beginning of Caritas, and throughout our history, we have crossed the line repeatedly, without hesitation, refusing to be silent amidst consequences of threats and attacks, in order to stay true to Our Lady. This is why you do not read or hear out of other Medjugorje promoters what you read and hear from these writings. They, the others, stay in line, not to be maligned and to remain in favor, and in the 'circle,' as referred to earlier. They do not want to be crushed for not being in alignment with what, the Medjugorje "authorities" have created, which is different from the Medjugorje "Our Lady created." A religious order did not call us or choose us or you; Our Lady did. Our Lady not only chose us, but Our Lady commissioned us, our mission—and—our "Way of Life," when She gave a direct message to us on May 31, 1995:

> **"…Get as many hearts as you can close to my heart and lead them to God, to a way of salvation."**

We are affirmed by Our Lady's own words to us, through Medjugorje visionary, Marija.

We have been built and continue to build on the foundation of Our Lady's messages and apparitions, and She has guided us to **a Way of Life in a New Time**. What is that life? A life of grace and of great trials. You can rest assured, if something is from and formed by Our Lady, you will not escape being persecuted, envied, and despised. While you build upon Our Lady's way, there are those who will build on their own foundation, on themselves. Many, to be in the circle of Medjugorje, place themselves under the powers that be, to be more accepted and strengthened in a higher position. What is our position? We put ourselves under everyone. We are at the bottom of the Medjugorje world, holding no position in the hierarchy of the Medjugorje structure. We were not chosen by anyone to be in the structure of authority of Medjugorje. We are outcasts, as Jesus was. But because we were chosen by Our Lady, our authority flows from Her. It is She, the Queen, who established what we do, and what we will do in the future.

Being chosen for what we do, does not put us above anyone else, rather we recognize who we are and what we are. What are we at Caritas? We testify that we are sinners, with faults, unworthy and generally no better than a worm. But there is one thing we are not. We are not suppressors and restrictors of Our Lady's apparitions, Our Lady's messages, and what Our Lady is asking us to do. We have put Our Lady's messages into the fabric of our lives. We are living Our Lady's "Way of Life," in a "New Time" that has not yet arrived. We live imperfectly, a perfect "Way of Life" that Our Lady has given us. What does that make us? Dirt! People of the dirt. That is what the great prophecy, the Corona Vision, revealed in the beginning of Medjugorje is!

The Agrarian Vs. The Urbanization of Life

The wonderful Corona Vision, described on the following pages, has been published by Caritas before, and I have spoken of it many times throughout the years. Now, in the time of the coronavirus, the meaning of the Corona Vision on Cross Mountain is defined further. Again, those who saw the vision did not know what it meant. It is explained in the book I wrote, It Ain't Gonna Happen: A Return to Truth™, released in 2010. The Corona Vision follows.

The Huge Vision, its Meaning and How it Applies to All Today

In Medjugorje, in the beginning days, people saw a strange occurrence on Cross Mountain. Some even saw it from afar. One of those who witnessed it was the visionary Marija's brother, Andrija. He said that he and others saw the whole of the sky over Cross Mountain, covered in what looked like a white veil, except one could see through it. Through the veil, they could see a small church with four or five houses around it. The four or five houses were surrounded with green fields. Then there was another church with four *or five houses around it, surrounded by green fields. And then another and another repeat of the scene. They were scattered across the sky, descending together, not one by one—floating down slowly, not in order, but all separated by open space. When Andrija was asked how many, he said hundreds of churches were surrounded the same way by a few houses. How long did the vision last? "Fifteen minutes." Did they just stay in the sky? Andrija repeated, "No, the little villages, churches were descending to the earth (on Cross Mountain) very slowly." Through the years, Andrija was asked about this several times, and what he thought it meant. He would say, **"It meant what I saw."***

Andrija was not the only one questioned about the vision. I spoke with another villager who collaborated the same scene. Andrija had no understanding as to its meaning except, *"It meant what I saw."* This huge vision covering the sky, the entire length of Cross Mountain, descending on the land, is how we are to be living <u>now</u>. By your choice or not by your choice, you will have to adapt to a way of life radically different from the modern way the world lives now. **Our future life will be the agrarian life vs. the urbanization of life.** Circumstances will decide which will be dominant. How? We have in our midst a foretelling event that easily leads us to reflect and discern that we are not living right. Look at what the coronavirus has resulted in: people breaking down into small groups, being more homebound, paralleling the vision of the churches surrounded by only a few houses. **This is a prophetic window of how the world will live in the future.**

One year ago, no one could have imagined in any way possible, how the world could so quickly be separated into little groups. It is amazing that in only a few months, we have been confronted with the breaking up of the population into small groups. This has compelled many not only to rethink how they live, but now influenced by a greater understanding of Our Lady's prophetic vision, is giving an impetus and opportunity for everyone to begin implementing a Way of Life in a New Time! Our Lady said:

September 2, 2011

> "...**Everything around you is passing and everything is falling apart, <u>only</u> the glory of God <u>remains</u>...**"

How many universities/schools, how many companies, how many institutions, how many banking systems, how many organizations under the appearance of good, are in reality not glorifying God today?

THE BIG BANG

BANG

BOOM

GONE, GONE, GONE

1. EVERY<u>THING</u>
2. AROUND
3. YOU
4. IS

5. PASSING
6. '<u>AND</u>'
7. **<u>EVERY</u>**THING
8. IS (not maybe, but '<u>is</u>')
9. FALLING
10. APART

What will be standing? What will hold together?

11. ONLY
12. THE
13. GLORY
14. OF
15. **GOD**
16. REMAINS

There are those who think the godless world, built on the sandy foundation of modernism, consumerism, and materialism, is going to stand against the wind of the Woman who is here, blowing away what is not of God's glory. The Woman of Revelation <u>is here</u> and it is She who has said Her sign is the wind. Our Lady said:

29

February 15, 1984

> **"The wind is my sign. I will come in the wind. When the wind blows, know that I am with you…When it is cold, you come to church…I am with you in the wind. Do not be afraid."**

Corona in Latin means crown. In reference to Our Lady's words about the devil, already stated, do you think Our Lady wasn't aware of the words of March 25, 2020, when She said:

> **"…satan is reigning and wants to destroy your lives…"**

Who is reigning, wears a crown. satan is king of the coronavirus, a virus that has gone viral worldwide. *For those who think the coronavirus is not of God, you are correct!* Medjugorje visionary, Marija, said the coronavirus is diabolical. Yet, it serves God's purpose in redirecting the world to be ready for Our Lady's plans to be fulfilled. One may think, "God would not purposely allow satan to do this, would He?" satan's participation in killing Christ on the Cross, was to satan's own horror, because it led to man's salvation. Likewise, the coronavirus is a forerunner; you might even say a shadow of the Secrets. The consequential results in regards to the coronavirus, give a sneak preview to what the consequential results of the Secrets will achieve, except on a much more massive scale. Our experience in these days has changed life across the world, a present prototype being endured as to what the Secrets will do. Just as the Secrets will do, the coronavirus is causing everything to fall apart. Again, Our Lady's words:

"…Everything <u>around</u> you is passing and <u>everything</u> is falling apart…"

- Look at the economy—it is endanger of falling apart.
- Look at the many non-essential and some essential jobs—falling apart.
- Look at the evil government school systems, teaching perverted doctrines including many private schools—falling apart.
- Look at the god of sports—falling apart.

- Look at the culture that has family members going in every different direction—falling apart.

And you can still list many more 'falling apart' statements.

Yes, after this present coronavirus trial, many things will resurrect, but this is only a preview of what is to come, in the real time and post time of the Secrets. When the real events of the Secrets begin to happen, and the great falling apart begins to unfold, there will not be a resurrection of what did not glorify God. Gone, Gone, Gone.

Who could have foreseen that the life of man, throughout the whole world, could change direction, practically overnight, by a tiny virus, and how prophetic it was that it was named Corona! But this is often how satan works as he wrestles to wear the crown, to be "king" over you. Our Lady has warned us of this:

October 2, 1992

"...Dear children, satan, in this time, wishes to act through small, small things,

dear children. Therefore pray! In this time, satan is strong and he desires to change your direction†; also my plan of peace, he wishes to destroy."*

If he can create so much chaos through, as Our Lady said, **"small, small things,"** what will happen when something comes much bigger than a **'small, small' virus**? This tiny virus disrupted life over the entire earth? What will happen when the Secret(s) are announced? It will be a shock of the senses, to change man's direction.

With the Secrets, many things will pass and fall apart, leaving in their trail, a purified culture of people, whose way of life will glorify God, under the Holy Corona of the Queen, the Queen of Peace! Purification is a causation to change the way one lives. The coronavirus has brought about the conditions for people to be homebound.

- Families are more together.

---

* † Our Lady said, the Croatian word, ***skrenuti***, which not only means to change your direction, but to divert your attention towards something else.

- People are not sitting in endless hours of traffic every day.
- Families are eating dinner together again.
- People are more often in contact with their families and friends.
- Neighbors are becoming more connected with each other.
- People are living life more in the moment.
- Large sectors of the work force are returning to the home.
- Children are being homeschooled by their parents.
- Children are happier, receiving their parent's undivided attention—which is pleasing to God.

Our Lady said on October 24, 1988:

> **"…pray for the young of the whole world, for the parents of the whole world so they know how to educate their children and how to lead them in life with good advice. Pray, dear children; the situation of the young is difficult…"**

While society is still sated in modernism, it is a step towards bringing families together, where life is focused on family again, the foundation for all other of society's structures.

The coronavirus has caused more reflection on changing one's future. Our Lady said on January 25, 1997:

"…reflect about your future…"

The coronavirus will be looked upon in the future as being a moment that caused a minimum impact upon our lives, especially compared to the impact the Secrets will make. Yet, the changes that happened when the coronavirus disrupted life on earth is an example of what is coming, and it is causing billions of people to think about reordering their lives, especially for those of us who are aware of the coming of the Secrets and the monumental, epic, massive reordered changes that will happen to life on earth. Our Lady said:

June 25, 2019

> **"…I am <u>preparing you</u> for the new time…that the Holy Spirit may '<u>work</u>' <u>through you</u> and <u>renew the face of the earth</u>…even though satan wants war and hatred…"**

It will take a lot of "<u>work</u>" through you to renew the face of the earth. We can define the change in the world's terminology. "The face of the earth" is about to get a make-over from an ugly face to a beautiful face—from satan's ugly face to Our Lady's beautiful face.

Our Lady is here to reorder the world that it may have peace.

Reorder to What?

The mammoth, descending Corona Vision on Cross Mountain is a picture of an agrarian world of very small communities. This is what the Community of Caritas is—a church surrounded by a few houses. No, *The Tabernacle of Our Lady's Messages* is not a church, but it has three private chapels, and the life of the community revolves around it. We live "church" together, being in communion together, in community, while still faithful to all the tenets of our faith. And what we have done, others can do, they can even build their own private chapel.

While the nations', and much of the world's, school systems are down, our children have not missed a day of school during the coronavirus! Our one-room schoolhouse, where we educate our children is still open. We are still working, full time, propagating Our Lady's messages for the world, while immersed in an agrarian life. Our community is together all day, every day, with our children, and we work, pray, carrying our crosses and joys as well. We grow much of our food together, play together, often eat together, work the mission together, and on and on. We are not social distancing ourselves from each other because we are a close-knit group on our own grounds. Our Lady, through Her apparitions, supported by Her messages, grew us, structured us, gave authority, brought us from the spiritual realm of the vision on Cross Mountain to a manifestation of a Way of Life into the physical realm of the Corona Vision. Our Lady called us to be a reality of the Corona Vision, that crowned Cross Mountain, to witness to others how to live a Way of Life in a New Time. Our Lady said:

January 1, 2011

"...help me realize my plans..."

Our Lady's plans <u>work</u>. We have witnessed it. All of our Medjugorje projects are moving forward, despite the loss of donations because of so many donors who are under hardship, and also because we are having some crisis repairs on our own grounds. Other than that, our lives are normal. We are doing our mailings across the world. Our press has not slowed down, but rather increased. Regarding our lives, hardly anything has changed.

Do you think that the Corona Vision on Cross Mountain is in your future? Do you have a choice? Yes, for the first question. No, for the second one. Now is the time to band together in small prayer groups, and evolve through prayer into communities. Medjugorje visionary, Ivan said:

"The prayer groups by us and in the world are the only answer to the call of the Holy Spirit. It will only be possible through prayer to save modern mankind from crime and sin."

Caritas began as a prayer group, with one family, who witnessed for Our Lady, and grew into a community. For those who follow in this path, it will not be easy. But the beauty of the life is what God will bring about, through four basic messages, not five, not three. Our Lady said:

June 6, 1988

> **"...I ask you to renew in yourselves the messages I have given to you. These are messages of prayer, peace, fasting, and penance...All of the other messages come <u>from these four basic ones</u>, but also live the other ones..."**

The four basic* messages of which all the other messages come from!

1. Prayer
2. Peace
3. Fasting
4. Penance

We have many people, who have told us when everything falls apart, they know where they are going. One person told us they have their camper filled to the ceiling with canned goods and food, ready to

---

* Some add the Mass with the four basic messages, thus saying there are the five basic messages. It is not so. First, the basic foundation had to be implemented to receive the grace of the Mass to be lived properly. Through the four basic messages. Our Lady's wisdom was to first prepare Her children to have the basics so as to understand the greatest prayer in the universe, the Holy Mass. **"...messages of prayer, peace, fasting and penance...'All' of the other messages 'come' from these four basic ones..."** 06/06/1988

come to Caritas. Others have told us the same. But we do not exist to become a city. The site of Caritas is a place of conversion. Tens of thousands have converted on this holy ground. It is a proven place of conversion. Hundreds of thousands that will grow into the millions across the world that have converted, many simply because of knowing what Our Lady has established here. Our Lady said:

July 25, 1999

> **"...I desire for you to comprehend that I want to realize here, not only a place of prayer but also a meeting of hearts..."**

As you have already read, we exist to be a reality of the prophetic Corona Vision given in 1981 — one private 'church,' that is the center of our life, *The Tabernacle of Our Lady's Messages,* surrounded by a few houses. You don't have to build a large private working chapel like ours. Ours is large because its purpose is to serve you and to propagate Our Lady's "Way" to the entire world. All you need is to take the messages seriously, birthing your little private chapel, with a few houses around it. Our Lady began in our home, with many apparitions over a span of 32 years, with many messages and things Our Lady did. Marija said on September 12, 2014, with about 80 people or so present in the home,* just after Our Lady appeared:

> *"I remember the first time when we were here (1988). Many people came with many flowers. After, so many people came to pray, overflowing from the Bedroom into the living room. The reality is that this house became…**this house become a chapel**, as Our Lady appeared in the living room, the sleeping room, Bedroom and other rooms."*

The home gave birth to *The Tabernacle of Our Lady's Messages*. We are not here in the valley of Caritas to save anyone from future events, rather we are to witness and teach you to begin to manifest

---

* In November, 1988, the Medjugorje visionary, Marija, stayed in the home of a Friend of Medjugorje and his wife for almost three months. Marija donated her kidney to her brother Andrija and the operation took place in a hospital in Birmingham, Alabama. During the three months that Marija was there, Our Lady continued to appear to her every day in the home. Whenever Marija comes back to Caritas, Our Lady appears to her in a Friend of Medjugorje's home. More details about the home will be found later on in this writing.

in the physical realm what was manifested in the spiritual realm, through a vision that crowned Cross Mountain. It is this that will save you from future events, when everything begins to fall apart. This is not said as a doom and gloom destructive prediction event, rather a renewing of the world to bring us to peace and to love the Creator, who is Father and Protector. But, much is up to us in this time of grace. If you do not do what you need to do, in the time of grace, you create a great problem. Our Lady said:

March 25, 1985

> "...I call on you—accept me, dear children, that it might go well with you..."

November 2, 2006

> "...Your time is a short time..."

But it is never too late. Our Lady says:

September 25, 2018

> "...I am calling you—it is not late..."

Living through this episode of the coronavirus pandemic, may make you afraid and tempted to panic as you read this writing. Instead, channel your whole life's effort into living the messages and act. Pray. Fast. Be calm. Be deliberate. Change what you can now. Prayer and fasting are important, coupled with tithing. Many skip over tithing. You give and God will give to you. In essence, tithing is giving something to yourself. Our whole mission is one of giving. Our Lady said:

December 25, 1992

> **"…your life does not belong to you, but is a gift with which you must bring joy to others and lead them to Eternal Life…"**

The community is a witness to you to show you how to become a small community village, how to reorder your lives, and how to change your direction. It is grounded in an agrarian way of life.

Our Lady said on January 2, 2000:

"...I, your Mother, am begging my children that they help me to realize what the Father has sent me for..."

Our Lady is here to 're'order the whole world, nation by nation; state by state; county by county; down to the small hamlets. Is this your future little village? The Corona Vision descending from the sky, crowning Cross Mountain, is a prefigurement of Medjugorje, before Medjugorje and its hamlets existed. In other words, the Corona Vision speaks of God's thought, represented in the vision across the sky, not yet a reality on earth. The Corona Vision began with churches surrounded by four, five or so houses descending from the sky. As mentioned already, there were hundreds of them. As they settled across Cross Mountain, something like a white veil was in front of the hamlets of which one could see through. Each little hamlet was a distance apart from each other and was surrounded by green spaces of pasture or fields.

While this vision speaks of what Medjugorje was like at the beginning of the apparitions, with the five small separate hamlets, it also speaks more of what the future world will look like, due to physical changes that will take place in the world, caused by the release of the Secrets. The Corona Vision is a crowning of a "Way of Life in a New Time;" a return to an agrarian life, close to God; a life God has ordained for man from the beginning. In the Book of Genesis, Adam was given words from God that he would no longer have the Garden of Eden, that was given to him by God, to live in and to eat from. After the Fall, Adam would be bound to the soil, a continued atonement, to always remind man that he depends upon God for his survival. Man was given a primary occupation, the only occupation ordained by God for man, with the words of Genesis:

> ***"You shall eat the plants of the field. In the sweat of your brow you shall eat bread till you return to the ground."***
> Genesis 3:18-19

Throughout history, every time man abandons his tie to the soil and goes towards urbanization, he becomes less dependent on God and, by doing so, man distances himself from God. This is where we are today. We are in an epic time, a time in which man has created a life so dependent on man, himself, and so independent from God, that man thinks of himself as god, even though he would not think that he thinks that way. Our Lady said:

October 1981

> **"...the West has made civilization progress, but without God, as if they were their own creators."**

Not only has Our Lady said man thinks he is his own creator, but Medjugorje visionary, Mirjana, confirms how far away man has strayed from what God ordained for us. Mirjana tells us* why the Secrets are coming:

---

* In a statement in response to a question posed by Fr. Petar in 1985.

"There never was an age such as this one, never before was God honored and respected less than now, never before have so few prayed to Him; everything seems to be more important than God. This is the reason why She cries so much. The number of unbelievers is becoming greater and greater. As they endeavor for a better life, to such people, God Himself is superfluous and dispensable."

The Secrets will correct and redirect man back to the path towards God. When man binds himself to the soil, God is not dispensable. Everything, every day, depends on God. Rain falls and seeds open and sprout by God's hand. Sun shines and creates temperatures for the seeds to grow. There are thousands of things that are dependent upon God in partnering with man to "subdue the earth." Urbanization separates man from God. That is why it is said that Abel was agrarian, a shepherd—the keeper of sheep, while Cain, after killing Abel, was cursed from the ground and left to create a city.

"Then Cain went away from the presence of the Lord….[Cain] built a city."
Genesis 4:16-17

We are in a time when man must humble himself, like Abel, or be humbled, like Cain. No wealth, no technology, no scientists, no inventions will be able to stop the deluge that is coming; just as in Noah's day, it could not be stopped. The difference today is the deluge, that is going to rain down, is a deluge of humility, crushing the towers of Babel we have built, and if you have not built one, we all have supported them for years. We helped to build them through our using them and also using our money to help support their system. The whole world is stuck in the system. We, at Caritas, have separated ourselves, as much as possible, and still transition away from the direction of the world. It is not an easy path to walk, but the fruit that comes from taking this path is a better life and a deeper connection to God. How can you get

THE CORONA VISION

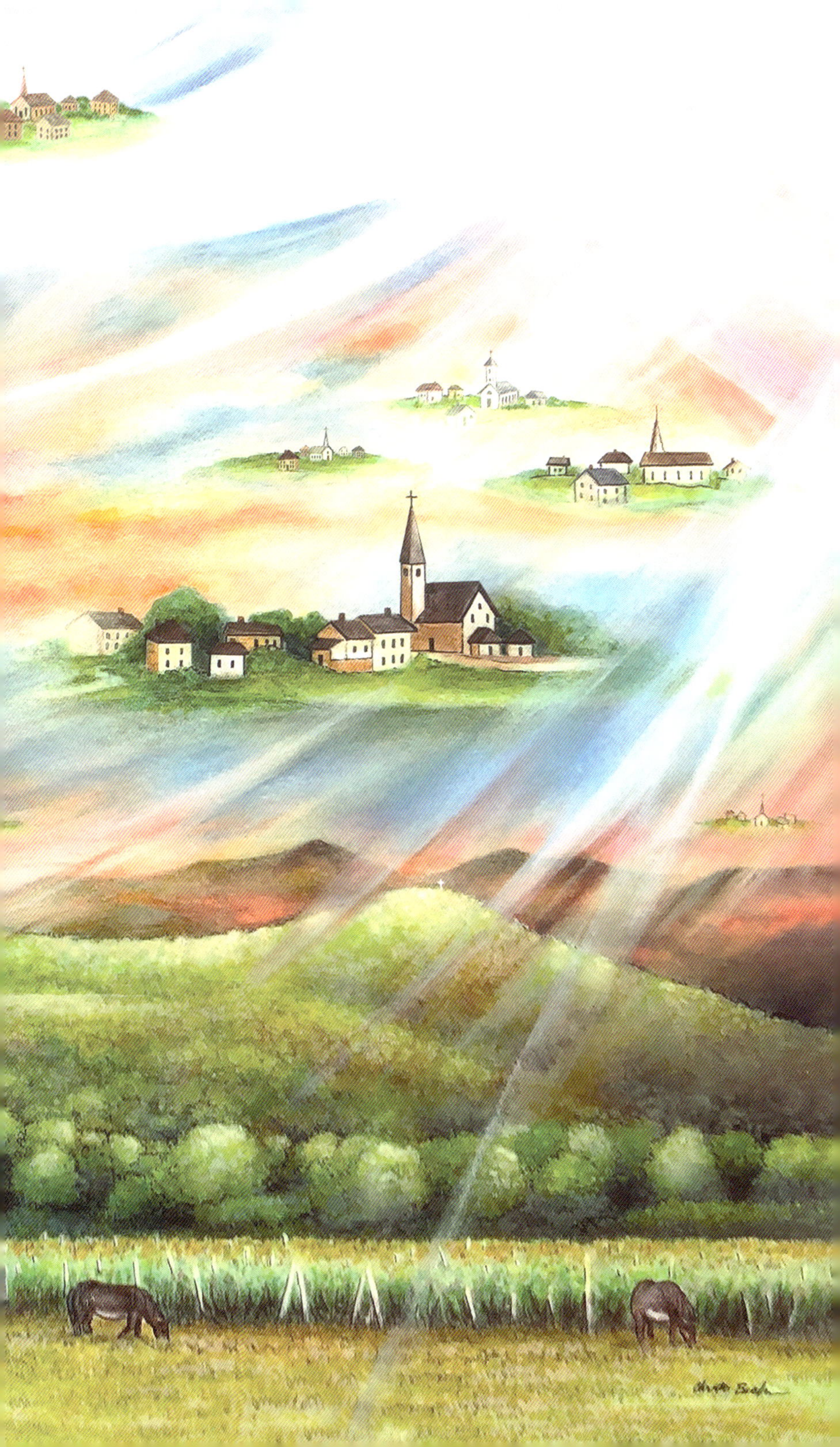

there more quickly? It was stated in a past writing, **"Man is never so close to God as when his knees are in the dirt."**

All that is written here is not to convey that man will have no other occupations outside of the agrarian life. Rather, what he is first to be occupied with is to be close to the soil to provide for himself, his food, even if more hours are spent in a trade or other occupations. Nor is this saying that you will be providing all your food yourself. One can reason that

"A Man is Never As Close to His Creator as When His Knees are in the Dirt." Written by a Friend of Medjugorje for the opening of the 2004-2005 School Year for *Our Lady of Victory's Little School House*.

in living in common, as the early Church did, life, with all its needs, is shared with each other, within a small community working together for the good of all.

St. Joseph was a carpenter, but he lived an agrarian life, drawing food from the soil, as all others did in Nazareth. My paternal grandfather, who we called Papa John, came to America from Italy in 1913 with nothing to his name. He went to work in the coal mines, but his primary support for a growing family of 10 was a garden, a milk cow, and always a couple of pigs. One may question, *"This is not practical. I am a doctor."* Well, be a doctor, but be agrarian! While we grandchildren called this small, but very strong man, "Papa John," most everyone else knew him and called him by another name. He was called "Doctor John." He was well-known for curing and healing many people because of his knowledge of herbs and medicine, knowledge that he brought with him to America from Italy. He was known to bring people back to health when nothing else worked. Additionally, he had a wealth of knowledge that he gained through working the soil and being raised in an agrarian way of life. His wife, Maddalena, whom we called "Mamalene," taught me only one lesson that I can remember. When I was eight or nine years old, she took a pitch fork from me, that I was trying to dig with, and showed me how to use it to break up the hard soil. There is an art to it and her agrarian knowledge helped me to do more digging with less effort, a principle I learned to apply to other areas of life.

The agrarian life not only feeds your body, it promotes health, supports your life, and also feeds the soul. Between the two, feeding the body and feeding the soul, the latter is primary. The coronavirus shut down public attendance to Mass across the whole entire world! Contemplate that if someone announced six months ago that six months from that moment, you and the whole world would not be going to Mass. You would have responded, by calling the person crazy or laughed it off. If the person predicted that you, with the whole world, would not be attending Mass even on Easter Sunday or taking Communion, and the world would be broadly shut down, again, you would have rejected the possibility of that happening. Yet, the impossible happened. What will you do now, knowing how quickly life can change in a blink? For decades, a persistent call has kept telling you that you must change the direction of your life. Have you taken that call seriously? Has your current situation, with the coronavirus, awakened you to the fact that you could have changed your life, but you didn't want to? The call, for 39 years, literally warned you. What of your future? Will you find yourself and your family in circumstances you could have avoided if you had listened to the call and acted on it?

August 25, 2013

> **"…I do not desire for you, dear children, to have to repent for everything that you could have done but did not want to…"**

Have you realized, suddenly, how suddenly life can change? What you were blind to before, do you now see? If you do not change your life after this coronavirus episode, what will be your disposition when the Secrets happen? Our Lady tells you what your disposition will be:

August 25, 1997

> **"…now you do not comprehend this grace, but soon a time will come when you will lament for these messages…"**

You will lament, not just for the messages, but because you did not put the messages into life by changing the way you lived.

Jesus came to the Eternal City of Jerusalem to save mankind. Our Lady came to an Eternal Place between the mountains.* Medjugorje is Eternal. Medjugorje is a New Jerusalem. Two Thousand (2000) years ago, Jerusalem served as Jesus' amphitheater for the salvation of man. Medjugorje serves Our Lady as an amphitheater to save mankind, calling man back to Her Son. Medjugorje is Eternal in the sense that billions in Heaven will be telling each

---

* The word Medjugorje literally means "between mountains."

other stories for Eternity that "my conversion" was through Our Lady, via Medjugorje. Our Lady said:

March 25, 1985

> **"…in a special way I have chosen this parish, one more dear to me than the others, in which I have gladly <u>remained</u> when the Almighty sent me…"**

As stated at the beginning, Medjugorje has no boundaries. It transcends the spiritual realm to the physical realm, intertwining both realms as one. No spot on the face of the earth has such a unique sacred touch with a porthole where Heaven can touch man and where man can touch Heaven. Our Lady, Herself, wedded Caritas' ground as a place of conversion, intertwining the spiritual and the physical with Medjugorje as one spot.

All above is scribed through
listening to the Woman of Revelation
Who is giving understanding to Revelation.

May 2, 2016

"....be apostles of the revelation…"

Everyone underestimates Medjugorje, including I, who write this. I testify that Medjugorje is bigger than anyone can understand, including the visionaries, and it will be prayerfully studied, pondered and meditated upon, realizing more, over time, its mysteries, until the end of time.

Revelation Chapter 21:2

> *"And I saw the Holy City, the New Jerusalem, coming down out of Heaven from God."*

Friend of Medjugorje

Friend of Medjugorje

Special Note:

There have been **204 apparitions** of Our Lady, through Medjugorje visionary, Marija, at Caritas. **148 apparitions** have been in the Bedroom of Apparitions, **1 apparition** when the Infant Jesus in the arms of Our Lady appeared in the loft of the home on Christmas Day with Our Lady, **6 apparitions** in the Living Room of the home of our founder and his wife, **45 apparitions** have been in the Field of Apparitions, **2 apparitions** have been at the Cross on Penitentiary Mountain, and **1 apparition** was in the writing office of a Friend of Medjugorje in *The Tabernacle of Our Lady's Messages.* These were accompanied with many messages from Our Lady. On several occasions, Our Lady appeared two times in one day. On other occasions, She came with angels. The apparitions took place throughout the years from November 1988 to the most recent ones in October 2019.

PostScript:

On page 54, I wrote about my paternal grandmother, Maddalena Colafrancesco, and the one lesson she taught me that was deeply etched in my memory. There was one other memory of a lesson given to me by her.

The second lesson Mamalene taught me occurred one day when she saw me itching and scratching a mosquito bite. She grabbed my arm and pressed her thumb nail really hard across the bite, then turned her thumb a quarter of a degree and did it again. A mark of the Cross of Jesus was imprinted over the mosquito bite and though it hurt while she was pressing hard on my flesh, after she was done, the itching stopped. As a young boy, I was amazed by this and I have passed it on to future generations. Today, Mamalene's remedy is used by all the kids in the Community. We've lost so many things of life by losing the agrarian life, a life of common sense connected to the spiritual. It fosters goodness, gratitude and humility.

    I wasn't going to include this story but something changed my mind right as this was going to press. When the picture of the village was found for the cover of this book, it perfectly captured the small village community as seen in the Corona Vision on top of Cross Mountain. It was almost too perfect and I asked if it was a real village. As it turned out, it is a real village located in the South Tyrol Valley in Italy. When I learned of the name of the church and the parish, I knew then that Our Lady's and Mamalene's hands were orchestrating things behind the scenes. Mamalene is the name that we called our Italian grandmother, but her given Christian name was Maddalena. To our joyful surprise, the picture of the Church in Italy in the tiny village is called:

"Chiesa di Santa Maddalena"

or in English

Church of St. Maddalena,

also known as St. Mary Magdalene.

Important

A Launching of Conversion

from a Friend of Medjugorje

Your support is absolutely critical for preparation of the Secrets to be ready for the onslaught of millions of conversions who will need the nourishment to grow in their conversions. Hundreds of thousands across the world have depended on the mission of Caritas to show them how to define Medjugorje into a way of life. There are almost a thousand writings, that cover a span of

cont'd on pg 64

The Tabernacle of Our Lady's Messages
Shelby County, Alabama, USA

The largest Medjugorje center in the world and Motherhouse of the Mission of Caritas of Birmingham. The 65,000 square foot 4-story building houses a state of the art graphic design department, printing presses, bookbinder, worldwide shipping department, broadcast RadioWAVE studio, Medjugorje.com offices, retreat center and BVM pilgrimage offices. Also, there are several buildings that provide support for Caritas' operations. A Friend of Medjugorje established, through the grace of Our Lady, the only infrastructure in the world dedicated solely for the archiving and propagation of Our Lady's messages to the corners of the world staffed with 50+ missionaries, plus one hundred Extended Community Missionaries across the United States—have dedicated their lives for this purpose.

34 years, concerning Our Lady's messages and how to live a "new life in a new time." Many who went to Medjugorje have lost their conversion, some no longer even believe in the apparitions of Medjugorje. Others grew tired, while others followed other interests. CONVERSION MUST BE FED or it will wilt and die.

How do you feed the world? The first task is prayer and fasting. Caritas has scheduled prayer times throughout the day to cover our work and our supporters.

No Medjugorje operation in the world has been established, outside of Caritas, to initiate, save, and preserve conversion on a massive scale, through God's grace. Caritas' existence, it's very life, is to "be ready" for the task. Yet, we are not where we need to be for the 12-month countdown, beginning June 25, 2020 to the 40th anniversary, June 25, 2021! One phenomenon of Medjugorje is conversion which <u>must</u> be nourished. Our Lady has given our mission a lot of food to feed millions when the Secrets begin unfolding. Yes, we have a book binder that can bind 50,000 books a day and other high production equipment, but what good is it, if there is not enough funds to pipe out the production the world will need? These conversions and materials are proven. We pose two serious questions to you.

1. You, who have followed the messages through writings, broadcasts, spiritual events from here—has God's grace, through Caritas, grown your conversion?

2. If yes for you, what of so many others who did not receive what you received through what you have read, heard or experienced on a pilgrimage here? Therefore, what good is conversion if the converted do not have the food to grow?

We currently are underfunded to meet the anticipated demands of souls that will need to be fed. We are asking for a serious commitment to help as Our Lady said, March 18, 2020:

> **"…sincere feelings, with good works; and help so that the world may change…"**

The Mission of Caritas of Birmingham, its base of operation, The *Tabernacle of Our Lady's Messages*, and the labor force for the mission, the Community of Caritas make-up a well-oiled machine, filled with prayer, but must be fueled by your good works of tithing or it comes to a standstill. To operate, without being crippled, to be ready for the 40th Anniversary of Our Lady's apparitions, June 25, 2021, we are asking you to commit—be one person out of 1,000 to tithe $50.00 per month. We must have 1,000—of you to be a special part of our Field Angel funding base. If you already are giving monthly, we plead with you to add $50.00 on top of your monthly tithe (we cannot afford to take away what you are presently giving). If you are not giving at all, please sign up to be one of the 1,000, $50.00-a-month-new Field Angels. This will give us a solid budget to run smoothly, and be more prepared on a daily basis, the operation of Caritas, allowing us to focus and produce more.

    This 1,000 $50.00-per-month request has been contemplated for the last two years. With the 2nd of the month apparitions ending on March 2, 2020, it is an absolute necessity to launch this increase to be ready for the secrets. All Christians are commanded Biblically to tithe 10% of one's income. If you are not tithing, you are losing your blessing that comes with giving. You **always** receive back. What is meant by "receive back?" In essence, tithing is giving something to yourself. The Bible tells us something profound about tithing:

Tobit Chapter 12:8-9

> *"…Prayer and fasting are good, but better than either is 'almsgiving' accompanied by righteousness…almsgiving saves one from death and expiates every sin…"*

Please call or write immediately for this urgent call. Our Field Angels keep the doors open for Caritas' daily operations. You are responsible for conversions every day and the strength to keep the converted "converted." Outside of Field Angels, we sometimes launch special specific spiritual projects of which we need funding for the expansion of the mission to be more broad spread in reaching the corners of the earth. Again, day to day, it is our Field Angels who are the foundation to keep the operations going. Thank you.

Again, please, as Our Lady said– with good works – help the world change. You can do that by contributing $50.00 per month or adding an additional $50.00 to what your gift already is for Our Lady through this mission.

In Grateful Thanksgiving
for Your Response,

Friend of Medjugorje

Friend of Medjugorje

P.S. For these 1,000 people who come forward to give $50 a month on this special Field Angel list, a Caritas Community member will be assigned to you personally and will make personal contact with you by phone and letter to answer questions or to keep you updated on major events or behind-the-scenes happenings that the general public would not know. In addition, if you call or write in for any specific prayer intention, your intention will be given to the Community member assigned to you who will be offering your intentions to Our Lady in their prayers and fasting. In other words, you will have your own personal prayer advocate. We make that commitment because it is that vital that we have the funding we need to be ready for the time of the secrets.

For more information on becoming a Field Angel or renewing your commitment call or write:

Caritas of Birmingham
c/o **Field Angels**
100 Our Lady Queen of Peace Drive
Sterrett, AL 35147 USA
(Outside the USA add 001) + **205-672-2000**

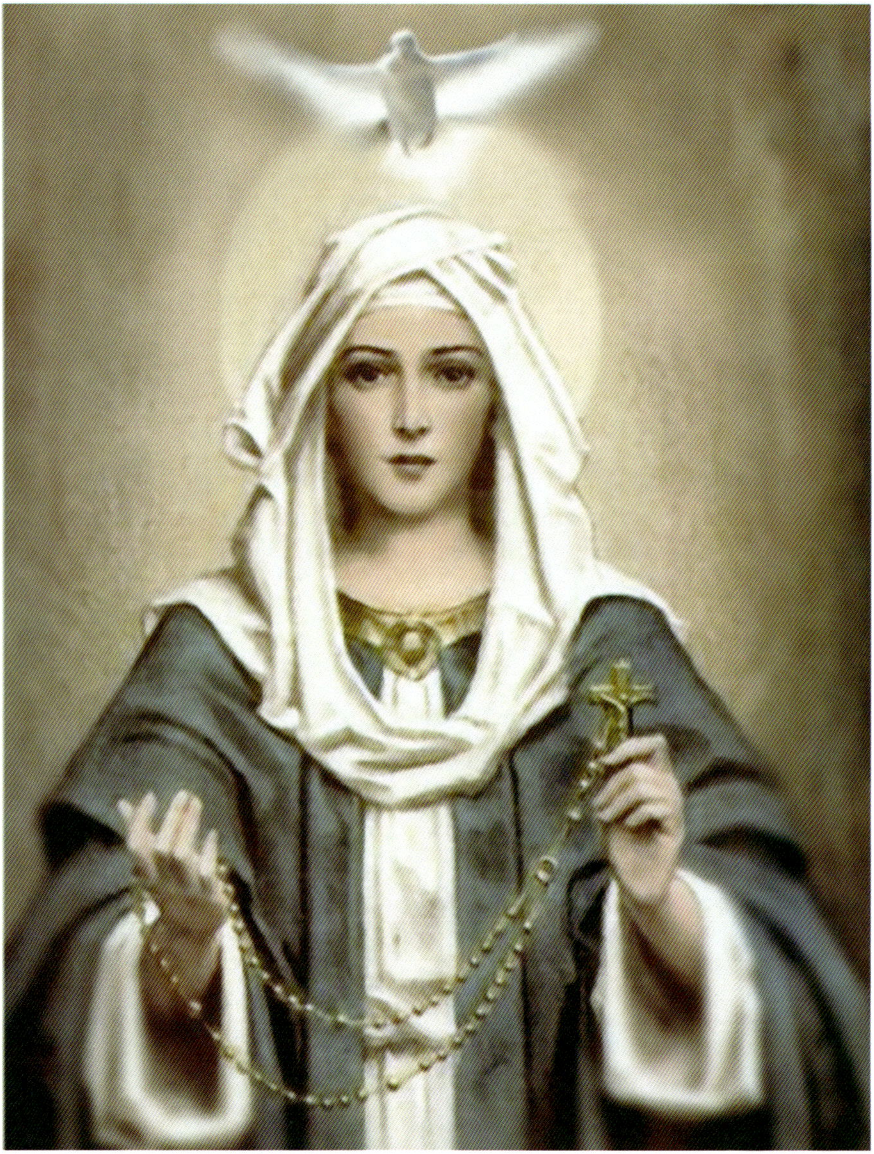

Spread This Book & Spread Our Lady!

With the coronavirus, turn a negative into a positive. Our Lady said:

May 25, 1991

> "...Dear children, I invite you to life and to change all the negative in you, so that it all turns into the positive and life..."

As the coronavirus spreads across the world, spread this book across the world. Because Caritas is a non-profit, The Corona Vision is priced for you to spread it far and wide. Give it to your family, friends, even your whole parish. It is a way to introduce many to Our Lady of Medjugorje.

To Order More Copies of The Corona Vision BF126

(Please add shipping and handling)

1 book=$4.00   10 books=$30.00 ($3 ea)   25 books=$50.00 ($2 ea)
50 books =$75.00 ($1.50 ea)   Case Pricing 100 Books=$100.00 ($1.00 ea)

CALL: Caritas of Birmingham
205-672-2000 ext. 315 twenty-four hours
or order on **mej.com** click on *"Shop Online"*
and click on *"Books by a Friend of Medjugorje"*

Glossary of What Does Our Lady Have to Say Concerning Two Different Worlds?

An Agrarian World vs An Urbanized World

"…this year how much seed I have sown…be my flower from that <u>seed</u>. Be my flower…" 09-08-2006

"…seed of faith may grow in your hearts; and may it grow into a joyful witness to others…" 01-25-2010

"…May prayer be for you like the seed that you will put in my heart…draw

"…Surrounded by material goods, how many times have you betrayed, denied and forgotten Him…do not deceive yourselves with worldly goods…" 11-02-2009

"…Do not live only for what is earthly and material…" 03-02-2016

"…Faith is being extinguished in many souls, and hearts are being

73

An Agrarian World VS An Urbanized World

you closer to God the Creator…" 01-25-2009

"…He (God) will give to you in abundance. As in springtime the earth opens to the seed and yields a hundredfold, so also your Heavenly Father will give to you in abundance…" 02-25-2006

"…if you live the messages, you are living the seed of holiness…." 10-25-1985

"…Every day I am sowing and am calling you to conversion, that you may be prayer, peace, love — the grain that by dying will give birth a hundredfold…" 08-25-2013

grasped by material things of the world…" 01-02-2019

"…your heart is filled with false glitters and false idols…" 08-02-2005

"…do not believe lying voices which speak to you about false things, false glitter…" 02-02-2018

"…Do not allow the false brightness that is surrounding you and being offered to you to deceive you. Do not allow satan to reign over you with the false peace and happiness…" 10-02-2003

"…Do not permit satan to open the paths of earthly happiness, the

An Agrarian World   VS   An Urbanized World

"…all the good that is in you may blossom and bear fruit one hundred fold…" 07-25-2011

"…Also nature extends signs of its love to you through the fruits which it gives you. Also, you, by my coming, have received an abundance of gifts and fruits…God will bless you and give to you a hundred-fold, if you trust in Him…" 09-25-2018

"…may a grain of hope and faith grow…" 01-25-2019

"…Your heart is like ploughed soil and it is ready to receive the fruit which will grow into what is good…Plant joy and

paths without my Son… they are false and last a short while…" 08-02-2010

"…It would be a good thing to give up television, because after seeing some programs, you are distracted and unable to pray…" 12-08-1981

"…you are in great temptation and danger because the world and material goods lead you into slavery…" 06-25-1989

"…Free yourselves from everything that binds you only to what is earthly…" 03-18-2016

"…do not look toward material things…" 12-05-1985

An Agrarian World   VS   An Urbanized World

"the fruit of joy will grow in your hearts for your good..." 01-25-2008

"...all which God has planted in your heart may keep on growing..." 04-25-1987

"...satan wishes to take advantage of the yield of your vineyards..." 08-29-1985

"...thank you for dedicating all your hard work to God even now when He is testing you through the grapes you are picking..." 10-11-1984

"...Prayer can teach you how to blossom..." 10-20-1984

"I desire that you be a

"...you are absorbed with material things, but in the material you lose everything that God wishes to give you. Don't be absorbed with material things..." 04-17-1986

"...realize that all the earthly things are not important for you..." 11-06-1986

"...Go on the streets of the city, count those who glorify God and those who offend Him. God can no longer endure that." 11-06-1982

"...See where satan wants to pull you into sin and slavery..." 01-25-2016

An Agrarian World   VS   An Urbanized World

flower, which blossoms for Jesus at Christmas, a flower which does not cease to bloom when Christmas has passed. I wish that you have a shepherd's heart for Jesus." 1983

"…I invite you to decide for life which blossomed through the Resurrection…" 03-25-1996

"…His words are the seed of eternal life. Sown in good souls they bring numerous fruits…" 09-02-2017

"…He is speaking to you with words of life and is sowing love in open hearts…" 10-02-2011

"…I may help you to be-

"…your spirit is weak and tired from all worldly things…be firm in this battle against evil…" 04-25-2015

"…The world and worldly temptations are testing you…" 02-25-2018

"…Sin is pulling you towards worldly things…you are struggling and spending your energies in the battle with the good and the evil that are in you…" 02-25-2013

"…Everything that is around you…leads you towards worldly things…" 11-25-2011

"…I am looking at your

An Agrarian World   VS   An Urbanized World

come a seed of the future, a seed that will grow into a firm tree and spread its branches throughout the world. For you to become a seed of the future, a seed of love…" 12-02-2011

"…permit the true faith to take root…" 09-08-1981

"…keep rooting them into your hearts…" 11-22-1984

"…you become more rooted in faith…" 06-25-1991

"…May the roots of your faith be prayer and hope in eternal life…" 02-25-2017

"You must begin to work in your hearts as you work in the field…" 04-15-1985

"…Do not forget that you ceaseless wandering and how lost you are…you would be able to recognize and to admit everything that does not permit you to get to know the love of the Heavenly Father…rise above the human way of thinking…" 12-02-2013

"…My Motherly Heart suffers tremendously as I look at my children who persistently put what is human before what is of God…They are walking to eternal perdition…" 03-02-2011

"…Free yourselves from everything that binds you

An Agrarian World   VS   An Urbanized World

are passing like a flower in a field…" 01-25-2007

"…The work in the fields is over…" 11-01-1984

"…begin to work in your hearts as you are working in the fields…" 04-25-1985

"…Let prayer…be your everyday food in a special way when your work in the fields is so wearing you out that you cannot pray with the heart…" 05-30-1985

"…Today I call you to start working on your hearts. Now that all the work in the fields is over…" 10-17-1985

"…this season is especially for you…When it

only to what is earthly…" 03-18-2016

"…Wandering in darkness, you even imagine God Himself according to yourselves…" 02-02-2011

"…do not permit the superficial brilliance of this world to mislead you: materialism, jealousy, arrogance. Do not permit the light of the world to mislead you…" 05-02-2016

"…Leave the passing things of this world of materialism—all that distances you from my Son…" 07-05-2019

"In Medjugorje, many have begun well, but they

An Agrarian World   VS   An Urbanized World

was summer, you saw that you had a lot of work. Now you don't have work in the fields, work on your own self…" 11-21-1985

"…You are so flooded by earthly concerns, you do not even feel that spring is at the threshold…As nature fights in silence for new life, also you are called to open yourselves in prayer to God, in Whom you will find peace and warmth of the spring sun in your hearts…" 02-25-2020

"I wish that you would become like a flower in the spring…" 06-11-1984

"…open your hearts to God like the spring flow-have turned toward material goods, and they forget the only good…" 11-18-1983

"…Renounce all passions and all inordinate desires. Avoid television, particularly evil programs, excessive sports, the unreasonable enjoyment of food and drink, alcohol, tobacco, etc…" 06-16-1983

"…satan, too does not sleep and through modernism diverts you and leads you to his way…" 05-25-2010

"…pray and fight against temptation and all the evil plans which the devil offers you through modernism…" 03-25-2015

An Agrarian World VS An Urbanized World

ers which crave for the sun…that He will always give abundant gifts to your hearts…" 01-31-1985

"…In this time of spring, when everything is awakening from the winter sleep…" 03-25-2009

"…Rejoice with me in this time of spring when all nature is awakening…"04-25-2002

"…I call all of you to grow in God's love as a flower which feels the warm rays of spring…" 04-25-2008

"…this time of spring moves you to a new life, to a renewal…" 03-25-2017

"…pray and fight against materialism, modernism and egoism, which the world offers to you…" 01-25-2017

"…there is much sin and many things that are evil…" 08-25-1992

"…the West has made civilization progress, but without God, as if they were their own creators…" 10-1981

"…due to the spirit of consumerism, one forgets what it means to love and to cherish true values… Do not let satan attract you through material things…" 03-25-1996

An Agrarian World   VS   An Urbanized World

"...offer to my Son Jesus for the coming of a new time - a time of spring..." 10-25-2000

"...your life is fleeting like the spring flower which today is wondrously beautiful, but tomorrow has vanished..." 03-25-1988

"...Let the family be a harmonious flower that I wish to give to Jesus...I wish that the fruits in the family be seen one day. Only that way shall I give all, like petals, as a gift to Jesus..." 05-01-1986

"...I desire to call you to grow in love. A flower is not able to grow normally without water. So also, you... are not able to grow without God's blessing....seek His

An Agrarian World   VS   An Urbanized World

blessing so you will grow normally…" 04-10-1986

"…When you pray, you are much more beautiful, like flowers which, after the snow, show all their beauty and all their colors become indescribable…open your inner self to the Lord so that He makes of you a harmonious and beautiful flower for Paradise…" 12-18-1986

"…I invite you to open the door of your heart to Jesus as the flower opens itself to the sun…" 01-25-1995

"…open yourselves to God as a flower opens itself to the rays of the morning sun…" 04-25-1998

An Agrarian World   VS   An Urbanized World

"…may your heart…open towards God as a flower opens towards the warmth of the sun…" 04-25-2012

"…keep praying until your heart opens in faith as a flower opens to the warm rays of the sun…" 04-25-2013

"…Open your hearts to the grace which God is giving you through me, as a flower that opens to the warm rays of the sun…" 04-25-2014

"…I will gather your prayers as flowers from the most beautiful garden…Be gardens of the most beautiful flowers…" 09-02-2018

"…Through the Holy Spirit you will become a spring of

An Agrarian World   VS   An Urbanized World

God's love…all those thirsting for the love and peace of my Son, will drink from this spring…" 04-02-2014

"…Drink out of my hands. My hands are offering to you my Son who is the spring of clear water…I may lead them to the spring of the clear water…" 10-02-2014

"…drinking from the spring of the words of my Son…" 07-02-2019

"Each member of the group is like a flower, and if someone tries to crush you, you will grow and will try to grow even more. If someone crushes you a little, you will recover. And if someone pulls a petal, continue

An Agrarian World   VS An Urbanized World

to grow as though you were complete." 06-21-1984

"…May each message be for you a new growth. Take this message into your life, in this way you will grow in life…"
02-17-1989

"…through prayer your faith grows and love is born…"
08-02-2013

"…You see…how nature is opening herself and is giving life and fruits…" 05-25-1989

"…Nature is awakening and on the trees the first buds are seen which will bring most beautiful flowers and fruit…"
02-25-2011

"…give glory to God the Creator in the colors of na-

An Agrarian World   VS   An Urbanized World

ture. He speaks to you also through the smallest flower about His beauty and the depth of love with which He has created you…may prayer flow from your hearts like fresh water from a spring. May the wheat fields speak to you about the mercy of God towards every creature…" 08-25-1999

"…As nature renews itself for a new life, you also are called to conversion…In nature seek God who created you, because nature speaks and fights for life and not for death…" 03-25-2019

"…Nature, in this way, is renewed and refreshed. For the beauty of nature, a daily

An Agrarian World   VS   An Urbanized World

renewal and refreshment is necessary. Prayer refreshes man in the same way…"

Follow up to January 27, 1986 Message

"…awaken your hearts to love. Go into nature and look how nature is awakening…" 04-25-1993

"…As nature gives the most beautiful colors of the year, I also call you to witness with your life…so that the flame of love for the Most High may sprout in their hearts…"
04-25-2011

"…When in nature you look at the richness of the colors which the Most High gives you, open your heart and pray with gratitude…" 09-25-2012

An Agrarian World   VS An Urbanized World

"…In this time of grace, when nature also prepares to give the most beautiful colors of the year…open your hearts to God the Creator for Him to transform and mould you in His image, so that all the good which has fallen asleep in your hearts may awaken to a new life…"

02-25-2010

"…I am calling you to a new life…transform you in this time of grace and, like nature, you will be born into a new life in God's love…"

02-25-2019

"…give thanks to God in your heart for all the graces which He gives you, also through the signs and col-

An Agrarian World   VS   An Urbanized World

ors that are in nature…"

08-25-2003

"…Today I invite you to go into nature because there you will meet God the Creator…God is great and His love for every creature is great…In the goodness and the love of God the Creator, I am also with you as a gift…"

10-25-1995

"…Every second of prayer is like a drop of dew in the morning which refreshes fully each flower, each blade of grass and the earth… How the scenery is beautiful when we look at nature in the morning in all its freshness…Man can thus become a really fresh flower for God.

An Agrarian World   VS   An Urbanized World

You see how drops of dew stay long on flowers until the first rays of sun come…"

01-27-1986

"…Bring it to all creation, so that all creation will know peace…" 12-25-1988

"…thank God, the Creator, even for little things…"

08-29-1988

"…open yourselves to God the Creator…" 02-25-1997

"…rejoice in God the Creator because He has created you so wonderfully. Pray that your life be a joyful thanksgiving, which flows out of your heart like a river of joy…" 08-25-1988

"…God has permitted me

An Agrarian World   VS An Urbanized World

to lead you towards holiness and a simple life—that in little things you discover God the Creator…" 11-25-2016

"…I call you to be love where there is hatred and food where there is hunger. Open your hearts…and let your hands be extended and generous so that, through you, every creature may thank God the Creator…" 9-25-2004

"…Today I call you to have your life be connected with God the Creator…" 04-25-1997

"…I call you to love first God, the Creator of your lives…" 11-25-1992

"…the world is in need of

An Agrarian World   VS   An Urbanized World

healing of faith in God the Creator…" 03-25-1997

"…open yourselves to God the Creator, so that He changes you…I wish to renew the world. Comprehend…that you are today the salt of the earth and the light of the world…" 10-25-1996

"…look at God's creatures which He has given to you in beauty and humility…" 02-25-2018

"It is raining at this time, and you say: 'It is not reasonable to go to church in this slush. Why is it raining so much?' Do not ever speak like that. You have not ceased to pray so that God may send you rain which makes the

An Agrarian World   VS An Urbanized World

earth rich. Then do not turn against the blessing from God…" 02-01-1984

"…there is only one spring from which you can drink…"
10-02-2019

"…I am asking for your roses of prayer…This is my dearest prayer which fills my heart with the most beautiful scent of roses…" 01-02-2017

"…give the gift of the rosary, the roses which I love so much. My roses are your prayers pronounced with the heart and not only recited with the lips. My roses are your acts of prayer, faith and love. When my Son was little, he said to me that my children would be numerous

An Agrarian World   VS   An Urbanized World

and that they would bring me many roses. I did not comprehend Him. Now I know that you are those children who are bringing me roses when, above all you love my Son, when you pray with the heart, when you help the poorest. Those are my roses…" 12-02-2017

"…so that hope, peace and love, which only God gives, may grow in your heart…" 08-25-2016

"…This time is for you…a time of silence and prayer… in the warmth of your heart, may a grain of hope and faith grow…" 01-25-2019

"…pray that faith may grow

An Agrarian World   VS   An Urbanized World

day by day in your hearts..."

01-25-2012

"...He was growing up alongside me..." 08-02-2014

"...I am calling you to adore my Son so that your soul may grow..." 11-02-2014

"...The fruit of peace is love and the fruit of love is forgiveness..." 01-25-1996

"...The fruit of prayer will be seen on the faces of the people who have decided for God and His Kingdom..."

12-25-2013

"...May prayer be a balm to your soul, because the fruit of prayer is joy, giving and witnessing God to others..."

10-25-2019

An Agrarian World   VS   An Urbanized World

"…I…encourage you not to give up from what is good, because the fruits are seen and heard of afar…"

11-25-2018

"…By looking at the fruits, your heart fills with joy and gratitude to God for everything He does in your life…"

10-25-2002

"…thank God for all the graces which God has given you. For all the fruits thank the Lord and glorify Him!…"

10-03-1985

"…The more you open yourselves, the more you receive the fruits…" 03-06-1986

"…work more on your personal conversion so that your

An Agrarian World   VS   An Urbanized World

witness may be fruitful…" 09-25-2010

"…An impure heart cannot give the fruit of love and unity…" 07-02-2011

"…Then, you will be my apostles who everywhere around them spread the fruits of God's love…" 01-02-2014

"…Faith is a most wonderful mystery which is kept in the heart. It is between the Heavenly Father and all of His children; it is recognized by the fruits…" 12-02-2018

"…But along with prayer and fasting you will be fruits, my flowers…" 08-02-2019

"…Find peace in nature and you will discover God the

An Agrarian World   VS   An Urbanized World

Creator Whom you will be able to give thanks to for all creatures…" 07-25-2001

"…You have no excuse to work more because nature still lies in deep sleep…" 01-25-1999

"…God wants to save you and sends you messages through men, nature and so many things which can only help you to understand that you must change the direction of your life…" 03-25-1990

"…Abandon yourselves to Jesus and only in that way will you be the sheep that follow their shepherd." 03-01-1987

"…permit God to lead you as a shepherd leads his flock…" 11-25-2006

Feedbacks Received from those who have read
<u>The Corona Vision:</u>

"Brilliant piece!"

Michael
Glasgow, United Kingdom

"Spot on! I cannot stress enough how much better it is to choose this 'Way of Life in a New Time' rather than be forced into it. Our Lady has gently guided us the entire way because we bravely said 'yes' to Her call. We went all in about 6 years ago, slowly. We would have never chosen an agrarian way of life on our own. It is hard, it comes with much sacrifice, it is humbling. Caritas has witnessed that toiling the land, putting in long hours brings forth many graces. We have felt so much peace throughout this man-made pandemic. Our everyday lives have continued as normal. We feel so blessed to raise our children in Our Lady's time. We are using this time to test what we have in place; to deepen our prayer, fasting & penance, re-evaluate our needs. Thank you for your love and sacrifice. Listen to what a Friend of Medjugorje has to say and act in a way that works within your means, but ACT."

Megan
Caledonia, Mississippi

"Dear Friend of Medjugorje and Caritas Family, I live in Fairfield County right outside New York City. We all remember the words 'Ground Zero;' it is not by chance that New York City is being called the 'Ground Zero' for the Coronavirus here in the United States. Life here is at a standstill. The way of life that we lived here pre-virus is over. People are doing the best they can to deal with this situation, but panic is in the air and there is real fear on people's faces, now covered with masks. It's hard to imagine any good coming out of this, but God is an awesome God and He will have the final say. This is terrible for people to have to deal with, especially the children. They are watching their parents coping with something they don't understand and are frightened. This will have lasting implications not only on our way of life, but, on the psychological wellbeing for the children in the future. All Churches and places of Worship are empty at a time when we need them most. I do believe this is diabolical and satan's entry into the Church is paying dividends in this evil scheme. Now we know why Our Lady has asked time and time again to Pray, Pray, Pray, and to pray for our Shepherds. Thank you, Friend of Medjugorje for your wisdom, insight, and prophetic words in regards to bringing a human simplistic interpretation of the Messages. Thanks for this Writing, so needed at this time. I did not find it to be scary as others have. It is Truth, and if we heed the Messages and the times we are living through, then we have nothing to fear. God bless you and the Caritas family always."

Anthony
Monroe, Connecticut

"After listening Thursday, I was a little afraid to come back and finish reading. There are so many things out of our control. So much I pray for, even for my own conversion, family, financial…knowing your vision and yet being so far from it in our lives. After reading, I don't feel panic for not being 'there' yet. I feel a bit of peace, as God's Word says, 'all things' He will work for the good of those called according to His purpose. So…I continue to pray for, and abandon my intentions to, Our Lady's plans for all of us and God's plan of salvation, for me and through me. **YOUR KINGDOM COME! YOUR WILL BE DONE! ON EARTH AS IT IS IN HEAVEN!** God grant me the grace to be even the least of Her apostles! May Your Will be done DESPITE me! Come Holy Spirit come! I have felt in prayer, this image: I am limping, with crutches. I am pulling a large rosary, like a yoke. Within are my loved ones. Some are trying to get out. From the outside of the rosary, demons trying to pull them out. All I can do is pray!"

Clovis, California

"My change unfolded in every line allowable to me even though satan was present. he puts up the best defense to the victory of our Mothers' Immaculate heart. I love this place [website] and hope to give everyone a Thank You back to holy Mother for the past 5 years I confirmed myself to her and the past ten years she has past by in the wind with me in her heart. Thank you, Friend of Medjugorje, for your holy ethics, and commitment to [your] writings. I could not have been here in heaven without y'all and your prayers and purity and open hearts."

Tashena
Bronx, New York

"I am so blest to have read this message. I know difficult times have brought me closer to Our Lady and Jesus. Today I am encouraged to continue to be part of two prayer groups. I am blest to belong to a church which has mass every day on the internet so my husband and I can attend during these days of the virus. We become a part of the supper of the Lord. I also have a chapel which I spend time daily saying the rosery and Divine Mercy. I do not say this in pride for as you have said. "We are nothing but dirt". But I am so blest to spend time in prayer and quiet. I was given the gift of going to Medjugorje and also your grounds in Alabama where I spent a few days in peace with Our Lady. My time on earth is short for me but as long as

I remain close to Jesus and Mary, I know the battle against the devil will be won. Thank you, my friend, for your example on how to live life to the fullest in the ways of Jesus and Mary."

Therese
Estero, Florida

"The words you speak are riveting. I see, my son sees, but so difficult to understand why some see and others do not. Could it be the seed itself? I pray for my whole family and in the isolation, I pray for all. I pray for strength, courage, fortitude to hear and listen to the Holy Spirit when He calls. I feel powerless at times, which I know what you are saying –that the devil is working overtime. I have hope that God's will be done on earth as it is in Heaven. I realize I am nothing and can do nothing but wait on the Lord. The title of the book you speak of and in search of is: 'Jesus is Coming.' I was shown this in a dream. 'Jesus is Coming.'"

Suzanne
Freehold, Select State

"Thank you for saying 'yes' to Mary's call. We have prayed for the conversion of America back to God for years and the conversion of families. I now ask myself if that was selfish to not include the entire world more in my nightly prayer. I have added to that prayer the conversion of the

world back to God, so we can all be prepared for the future. May God bless, protect and guide us to be apostles of His Word. God have mercy on us all and forgive our sins."

Paula
Sebastian, Florida

"A total conversion to God transforms our souls into a 'new Jerusalem' and I am joining my prayers and sacrifices with all of you for more conversions throughout the world. God bless you, a Friend of Medjugorje, your family, the Community and all the followers of Medjugorje messages."

Feliza
Santa Rosa City, Philippines

"Thank you, A Friend of Medjugorje, for responding to our Lady's call, for accepting the big responsibility of being a 'modern-day prophet', and for sharing all your insights and wisdom with the world! I imagine Caritas, as Medjugorje was for me, 'a little heaven on earth,' and I want and would like that SO MUCH in my home/family! But, what can we/us do who live in large cities... with family members with different thoughts/ideals/ideas, with family spread all over, with neighbors with different faiths, with a fixed income/low-income resources, with no land to work/live off of, etc.,. I know and understand that it's part of the cross, but I honestly and truly want to be a modern-

day apostle of love for our Heavenly Mother and thus yearn for that 'little heaven on earth!' Once again, THANK YOU and may our Merciful Father continue to bless you, and may our Loving Mother continue to guide, protect and grace you!"

Vera

Modesto, California

"Love and Truth, what more can one ask for?"

Tony

Fort Laramie, Ohio

"Thank you again for such wonderful insight. Thank you also for inspiring/leading so many last Fall in the 9-day bread and water fast for 2020, "so that it may go well for you." After this last desolate Holy Week, we family of 9, are more committed to Wednesday and Friday bread and water fasts."

Monica

Caledonia, Minnesota

"The wisdom and insight of a Friend of Medjugorje is profound in this writing. The first three secrets are about the village of Medjugorje. But Medjugorje has been dispersed across the world, especially in and through Caritas followers. And Extended Community members are a part of that. We all have Medjugorje, we have it in our hearts,

and Our Lady needs us to be where we are to make an impact. I have given my life to Our Lady. She needs me to remain where I am, or she would have made a way for me to go to Caritas. I would go tomorrow if she made a clear way for me to go. While in Medjugorje on a Caritas pilgrimage we were told, take Medjugorje home with you, make your home Medjugorje. By taking Medjugorje home in our hearts, all of our homes will become like chapels, it is a challenge that only Our Lady can help to make possible, to live in humility, and to bring joy to others, to help lead them to Eternal life. This is the gift, pray to be worthy. Thank you, Caritas, for all the guidance."

Camilla
Alexander, Arkansas

"I thank the entire Caritas community for your example and dedication to living & spreading the messages of Our Lady. From the time I personally witnessed the Miracle of the Sun when I was there at Caritas in March of 2011, I became convinced that, not only is Our Lady appearing there and in Medjugorje, of course, but that you all have been especially chosen to help spread Her messages. You have been a great inspiration to me and a light to the world. While I have not (as yet) had the opportunity to travel to Medjugorje, I have been so very blessed to be present at Caritas during Her apparitions to Marija on 17 occasions!

So incredible! I could feel the graces coming from Our Lady! But I do pray that Marija or any of the visionaries may return again this year—I hope, I hope! Whatever the case, I would like to add the I, too, see that good things are happening as a result of the coronavirus, as you already stated; i.e., parents homeschooling their children! PTL! Forever grateful…"

Elizabeth
Nederland, Texas

"Another great writing. Thank you, Caritas for all you do."

Cleveland, Texas

"Thank you 'A Friend of Medjugorje' for not only living the messages of "Our Lady," but for sharing them with the World! I've read your books & heard Radio Wave, traveled to Caritas with busloads of people & went to Medjugorje twice with BVM Pilgrimage & my Heart yearns to go again!!! I can still see Marijana's eyes filled with tears asking our pilgrimage group "What took us so long" (to come)? (And that was in 2009)! My prayers are with the Visionaries, who do suffer for the Salvation of Souls! I will be praying & fasting for the coming Secrets! Your words, "A Friend of Medjugorje" are TRUTH and your life is proof! The world will surely lament when this time of Grace ends! Praying to have ears that hear & eyes that see!

Thank you again 'A Friend of Medjugorje' for Spreading Our Lady's Messages to the ends of the World!!!"

Connie
Harwood, Texas

"*When I first read about the 7 day fast for Mirjana and Fr. Petar, I hoped that we could all join in with them. It looks like we may be able to fast with them and help them be strong. Thank you, God our Father, God our Savior Jesus, and Mary our Mother for this privilege. This Coronavirus has been a blessing in disguise—so many commitments have been canceled, I find myself praying so much more and so many more times during my day. May God open so many eyes, ears and hearts during this life without busyness.*"

Pete
Carpinteria, California

"*He Who Has Ears To Hear, Let Him Hear! Thank you for this prophetic writing Amen!*"

JA
Okotoks, Canada

"*I have been reading and following Our Lady's words from the beginning. Please pray for all of us that hearts will be touched by Grace and begin to believe that Our Mother loves us so much and is trying to save us from certain physical and especially spiritual ruin. Thank you for guiding us and leading the way all of these years!"*

> *M. Raphael*
> *Drexel Hill, Pennsylvania*

"With this statement from this reading, "You can rest assured, if something is from and formed by Our Lady, you will not escape being persecuted, envied, and despised," I feel that not many realize how active satan is in our world, especially when it comes to Medjugorje. People are quick to attack the negatives, but sure do know how to ignore the positive. I know a Friend of Medjugorje has talked and wrote about it before, but our cellphones and technology are things we really need to distance ourselves from no matter how hard it is and it is hard. Thanks for this writing and Our Lady of Medjugorje, please pray for us and watch over us.

> *John*
> *Ashland, Ohio*

"'For you who fear My Name the sun of justice will rise with its healing rays!' [Malachi 3:20] 'Turn the hearts of fathers to their children and the hearts of children to their

fathers, least I come and strike the land with utter destruction.'" [Malachi 3:24]

Anna

Princeton

"Thank you so much for this clear, concise, beautiful summary of what you have been saying for decades. Others need to hear it now. We did what Our Lady asked and humbly work at conversion every day, all day. AND still, with Medjugorje knowledge, Daily Mass (before now), Rosary, fasting and penance, STILL we need so much help. Thank you and we pray for you all every day."

Your Wisconsin Field Angels
Katie
Grand Marsh, Wisconsin

"What shall I do being married to a man who will only believe the first 7 messages of Medjugorje because that is what the Church approves? He doesn't believe the rest. He came to Medjugorje with me (my 2nd, his first time) to protect me from the evil he thought was there. He makes fun of me when I want to buy silver as he has much faith in stocks, even when I warn him that is not true wealth and can disappear overnight. How can I convince him to make the changes we need to make if he doesn't believe Our Lady has been coming for 39 years? I must add that *he was a daily Mass attendee before the virus hit so he does believe in God. And wears a scapular but says we are not required to believe in Medjugorje. Please help me here!"*

Karen
Troy, New Hampshire

Response to Karen from Caritas of Birmingham:

"Karen, you are one of many women in the same type of situation. We can tell you what we have heard a Friend of Medjugorje repeat so many times through the years. Pray for God to guide your husband. Live love in every way. Live your state in life, as a wife, to perfection. And storm Heaven with prayers to get your husband here to Caritas for a visit. Our Lady will do the rest."

* * * * *

"Thank you, Caritas! The Corona Vision is a hope filled future. I am inspired to build a chapel as of today. I am thinking of starting out with just a shed sized dwelling that is isolated from modernism. I am inspired and thankful for the reminder of the 4 basic messages. I have fallen numerous times since my pilgrimage to Medjugorje and to Caritas but am forever grateful for your witness and encourage-

ment. God bless you and the community. Where we go one, we go all! Big Q, uniting us for the Victory!"

> Dominic
> Yellowknife,
> Northwest Territories — Canada

"I continue to try to follow the messages but also fail miserably. Now, I have "family" living with me and it is so hard. I have increased my prayer life, watch the Mass on TV, pray the Divine Mercy chaplet and just generally try to pray all day. I do love working outside in my gardens when the weather gets better. This year will be attempting to grow more vegetables. Less flowers this year. I love Our Lady and I know She is with me but I do feel I fail Her so much. I am attempting to express how I am doing but not articulating myself very well. Thank you, Friend of Medjugorje! Yes, my conversion happened big time in 2016 in Medjugorje. Praise God and Our Lady for inviting me there. Now, if I can just help Her to accomplish Her plans!! I am a Field Angel also but unfortunately because of new living arrangements and costs cannot increase my giving amount at this time. No need to publish this just wanted to express my gratitude!"

> Kris
> Ohio

"Sorry: in addition, I know Jesus and Mother Mary are with me. I was able to get to Mass twice on the Saturday before lock down of USA. So, I know They are with me in a special way, assisting me in this trial of 'family' with different views and practices. I am struggling but Jesus and Mother Mary are here. As is also, God the Father. I am trying to complete my mission that I do not really know what it is—except maybe to continue to pray for my "family" that they come to the light, not be dependent on technology etc. It is a trial/suffering, but I offer it all to God and try to do that, remember, every present moment. I am just so grateful for the Friend of Medjugorje for enlightening me further. I guess I need to read <u>It Ain't Gonna Happen</u> as I did not know of this vision. Maybe my little household myself and the 3 family members (niece, great-niece and great-nephew) will be, eventually, a homestead and Church!!"

Kris
Ohio

Endnotes

1 Medjugorje visionary, Marija, went before Our Lady on behalf of a seminarian and asked the question: *"Was it okay to read the book The Poem of the Man-God?"* Marija relayed that Our Lady affirmed it was okay by answering:

"One can read them."

Because a storm of opposition erupted as a result of the writings of Maria Valtorta's growing popularity, Caritas of Birmingham, Cardinal Ratzinger, and Pope John Paul II are recorded in Church history as a result of a verdict rendered concerning The Poem of the Man-God.

The historic Church ruling cleared the way for the faithful to read the volumes. The Church only required they not be declared supernatural. This is always the way the Church works. Time gave the historic decision to Caritas and time will further clarify. What is important now is the Church's approval for the faithful to read them, and that The Poem of the Man-God can be read, correlates to Our Lady's words, **"One can read them."**

SPECIAL STATEMENT

Caritas of Birmingham is not acting on behalf of the Catholic Church or placing its mission under the Church. Its mission is to reach all people of the earth. Its actions are outside of the Church done privately. It is further stated:

So as not to take for granted the credibility of the Medjugorje Apparitions, it is stated that the Medjugorje apparitions are not formally approved by the Catholic Church.

Medjugorje Status
April 25, 2020 A.D.

No attempt is intended to pre-empt the Church on the validity of the Medjugorje Apparitions. They are private revelation waiting the Church's final judgment[1]. In the interim, these private revelations **are** allowed by, and for, the faithful to have devotion to and to be spread legally by the Church. Devotion and the propagation of private revelations can be forbidden only **if** the private revelation is condemned because of anything it contains which contravenes faith and morals according to AAS 58 (1966) 1186 Congregation for the Doctrine of the Faith. Medjugorje has not been condemned nor found to have anything against faith or morals, therefore it is in the grace of the Church to be followed by the faithful. By the rite of Baptism one is commissioned and given the authority to evangelize. *"By Baptism they share in the priesthood of Christ, in His prophetic and royal mission."*[2] One does not need approval to promote or to have devotions to private revelations or to spread them when in conformity to AAS 58 (1966) 1186, as the call to evangelize is given when baptized. These apparitions have not been approved formally by the Church. Caritas of Birmingham, the Community of Caritas and all associated with it, realize and accept that the final authority regarding the Queen of Peace, Medjugorje and happenings related to the apparitions, rests with the Holy See in Rome. We at Caritas, willingly submit to that judgment. While having an amiable relationship with the Diocese of Birmingham and a friendly relationship with its bishop, Caritas of Birmingham as a lay mission is not officially connected to the Diocese of Birmingham, Alabama, just as is the Knights of Columbus.[3] The Diocese of Birmingham's official position on Caritas is neutral and holds us as Catholics in good standing.

1. The Church does not have to approve the apparitions. The Church can do as She did with the apparitions of Rue du Bac in Paris and the Miraculous Medal. The Church never approved these apparitions. She gave way to the people's widespread acceptance of the Miraculous Medal and thereby the Apparitions to St. Catherine. *Sensus Fidelium* (Latin, meaning "The Sense of the Faithful"), regarding Medjugorje, is that the "sense" of the people says that "Mary is here (Medjugorje)."
2. Catechism of the Catholic Church Second Edition.
3. The Knights of Columbus also are not officially under the Church, yet they are very Catholic. The Knights of Columbus was founded as a lay organization in 1882, with the basic Catholic beliefs. Each local council appeals to the local Ordinary to be the Chaplain. The Knights of Columbus is still a lay organization, and operates with its own autonomy.

Published with permission from SJP Lic. COB.

© 2020, S.J.P. Lic. C.O.B.

ISBN: 978-1-878909-72-5

Printed and bound in the United States of America.

©SJP International Copyright. All rights reserved including international rights. No part of this book may be reproduced or transmitted in any form or by any means, electronic or mechanical, including photocopying, recording, or by any information storage or retrieval system, without permission in writing from Caritas who is licensed to use the material. Caritas of Birmingham, 100 Our Lady Queen of Peace Drive, Sterrett, Alabama 35147 USA. None of the mailing lists of Caritas or its entities, including electronic mailing lists, etc., are for sale, nor is permission given to use them in anyway, by anyone. There are no exceptions. All civil, criminal and interstate violations of law apply.

To Order More Copies of The Corona Vision
CALL: Caritas of Birmingham
205-672-2000 ext. 315 twenty-four hours
or order on **mej.com** click on *"Shop Online"*
and click on *"Books by a Friend of Medjugorje"*